K-2 Math
Volume 2

© 2013 OnBoard Academics, Inc
Newburyport, MA 01950
800-596-3175
www.onboardacademics.com

ISBN: 978-1-939796-73-8

Table of Contents

Compare and Order Numbers

Key Vocabulary

compare

order

less than

greater than

Draw the red numbers onto the number line.

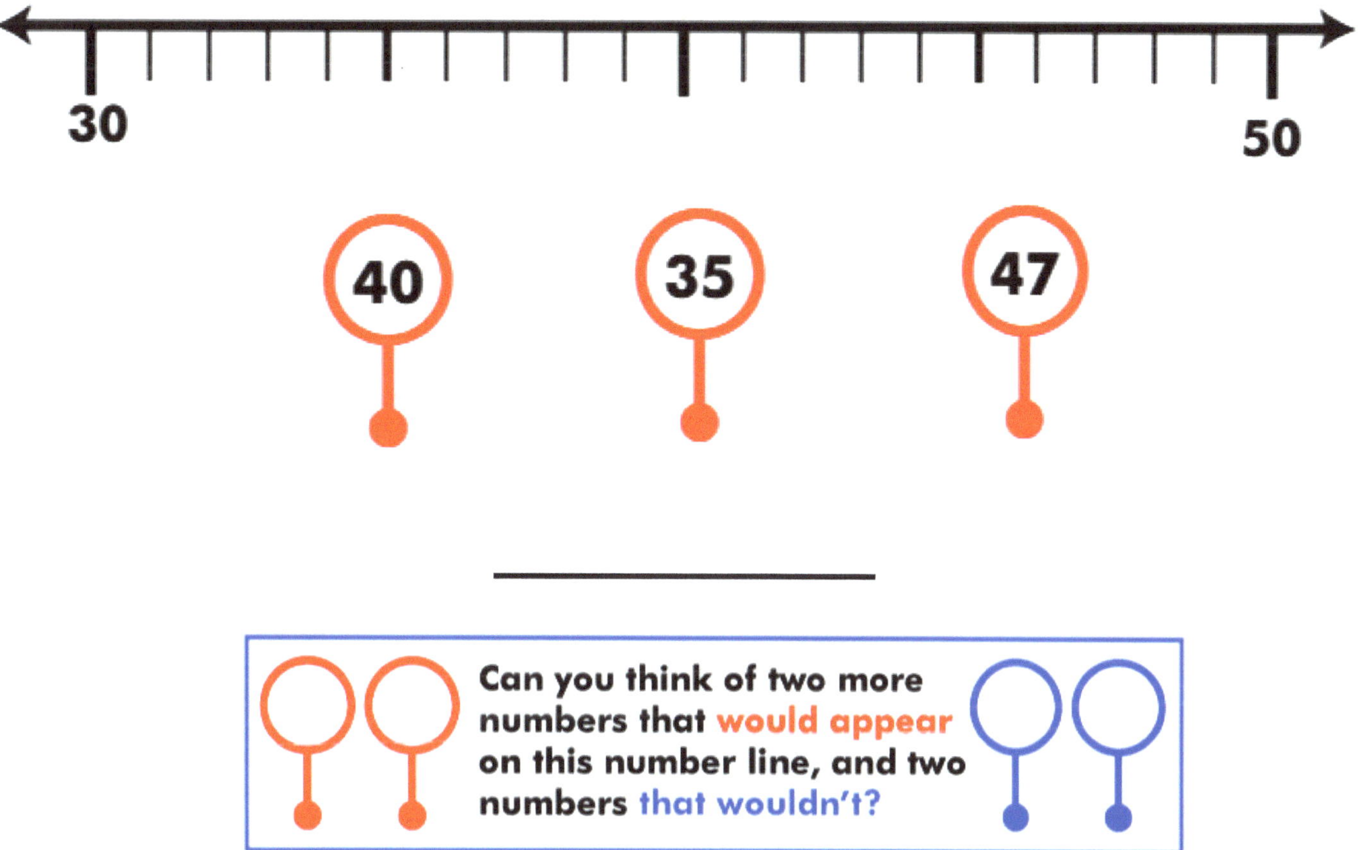

Write the numbers that would appear in the red circles and the numbers that wouldn't appear on the number line in the blue circles.

Comparing Numbers

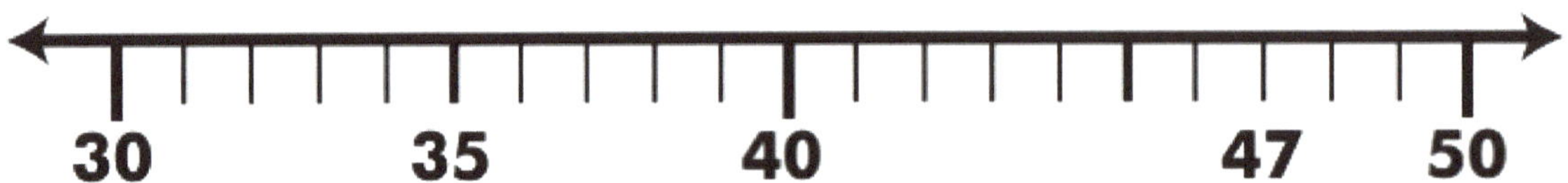

1 Which number is smaller, 40 or 47?

2 Which number is larger, 40 or 47?

3 Which number is smaller, 35 or 50?

4 Which number is larger, 35 or 50?

Locate these numbers on the number line. Draw them in their proper place.

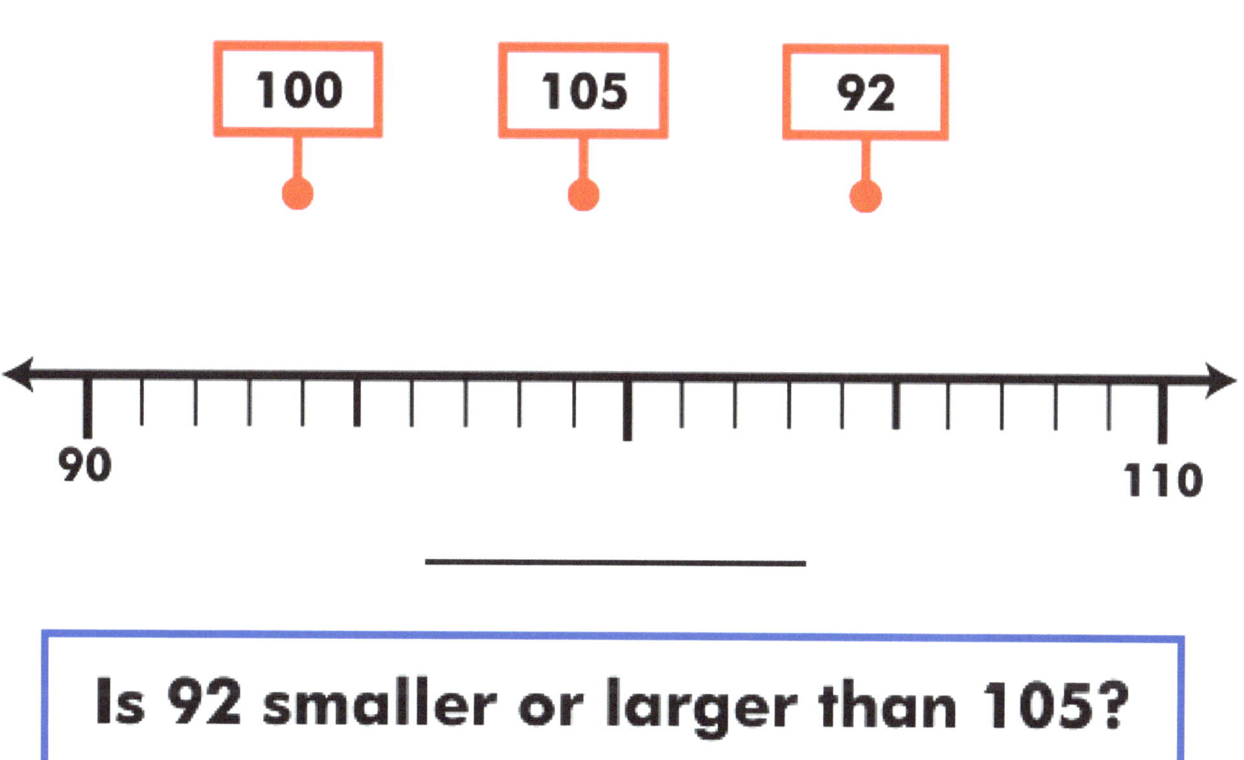

____

Is 92 smaller or larger than 105?

Write your answer here__________

____

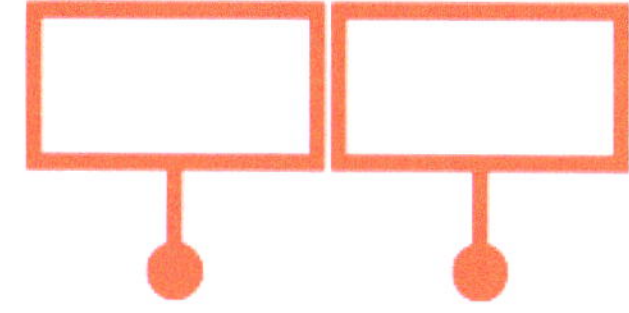

Can you think of two more numbers that would appear on this number line, and two numbers that wouldn't?

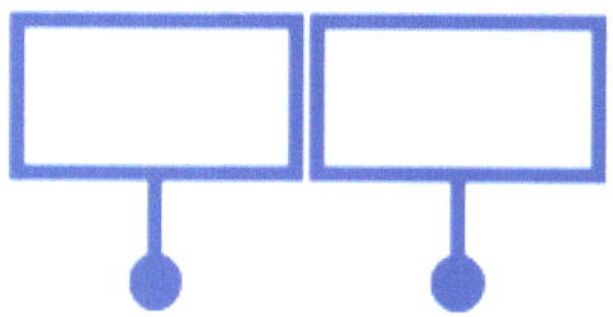

Write your answers in the red and blue boxes.

Locate 400 on the number line. Draw it in place. Now sort the numbers by writing into one of the boxes.

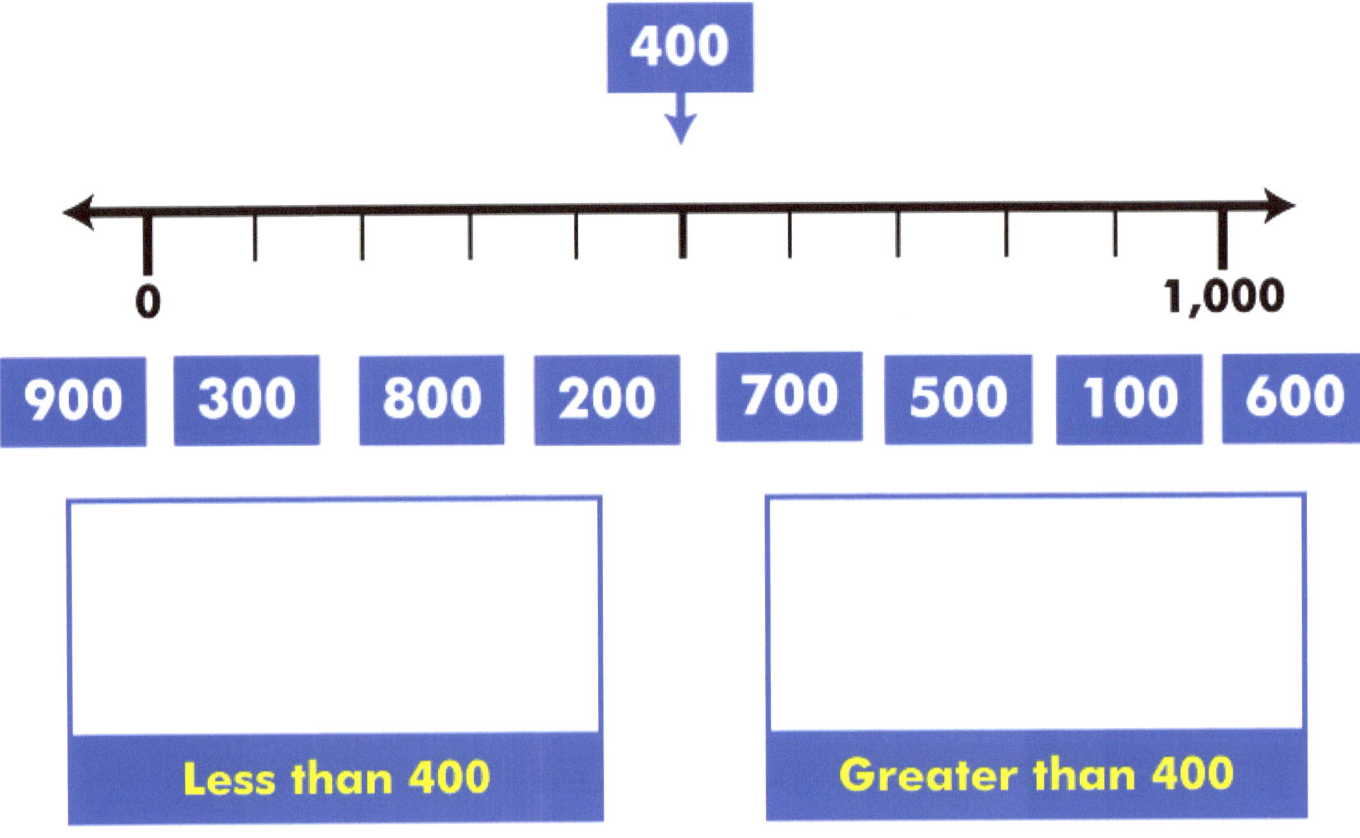

Less than, greater than, equal to

Study the illustration below. identify the numbers on the number line, reach the comparative statement and then look at the corresponding equation.

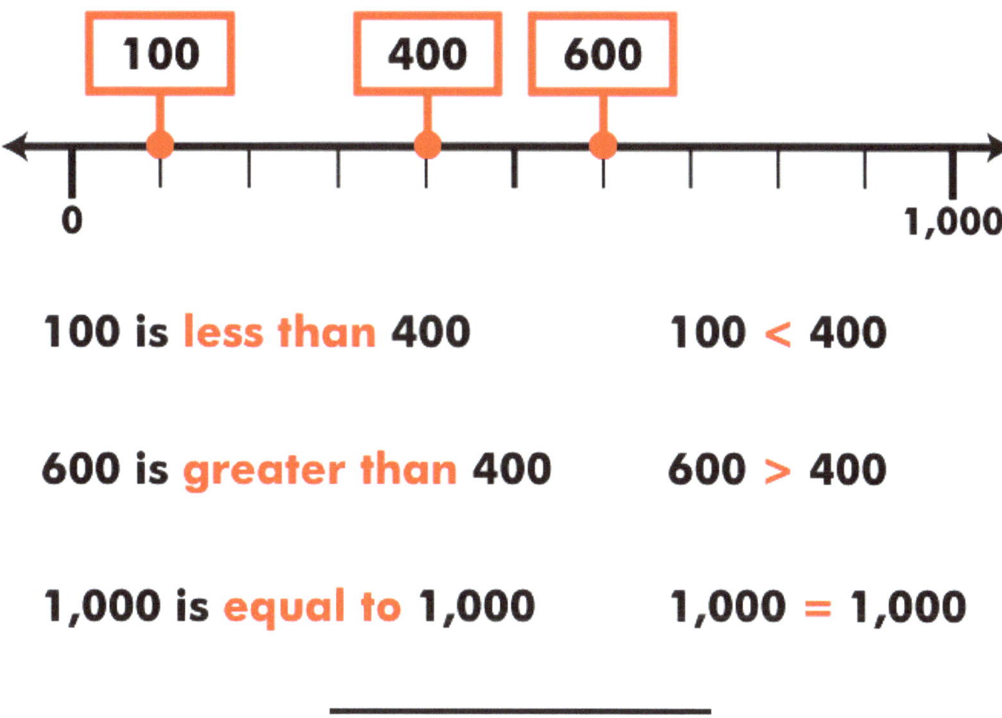

How can I remember this?

27 < 54

54 > 27

Look at the next page for a hint.

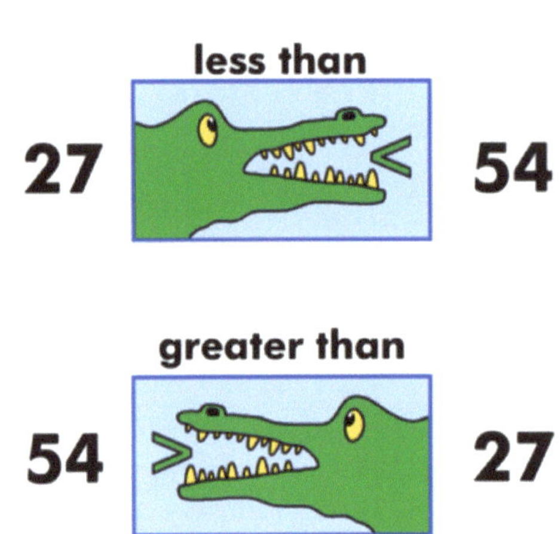

Less than, greater than and equal to practice.
Use the symbols below to complete the equation.

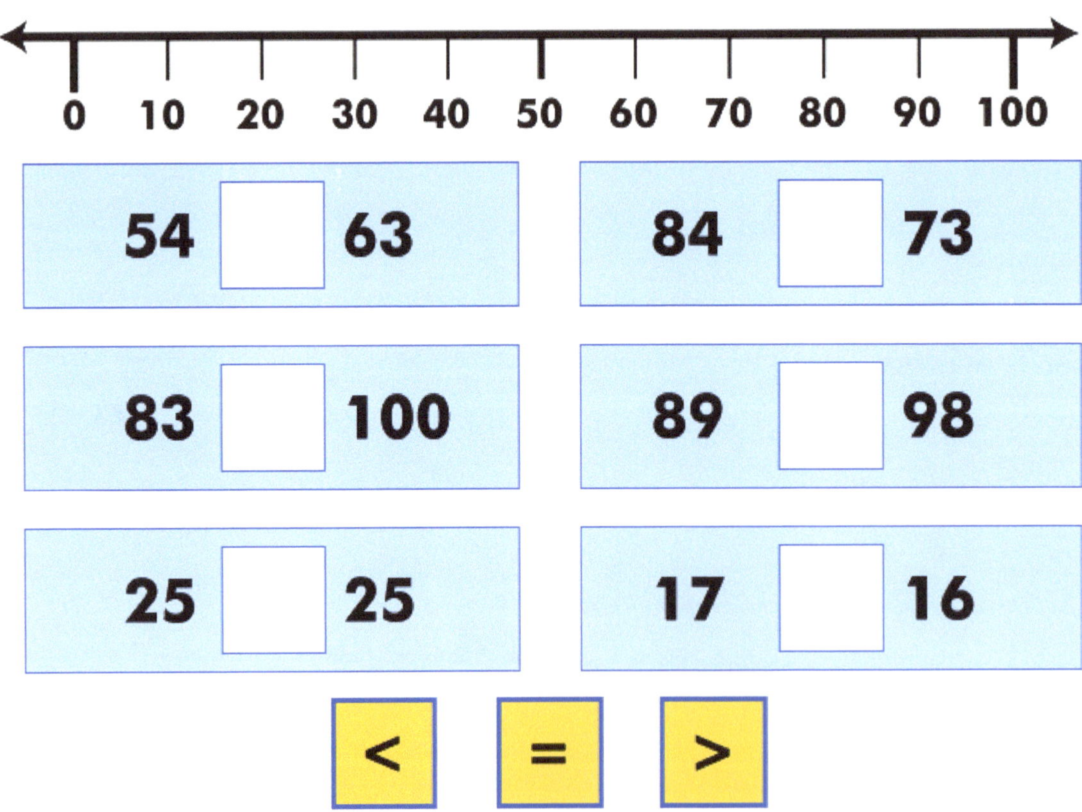

More practice, fill in the blanks.

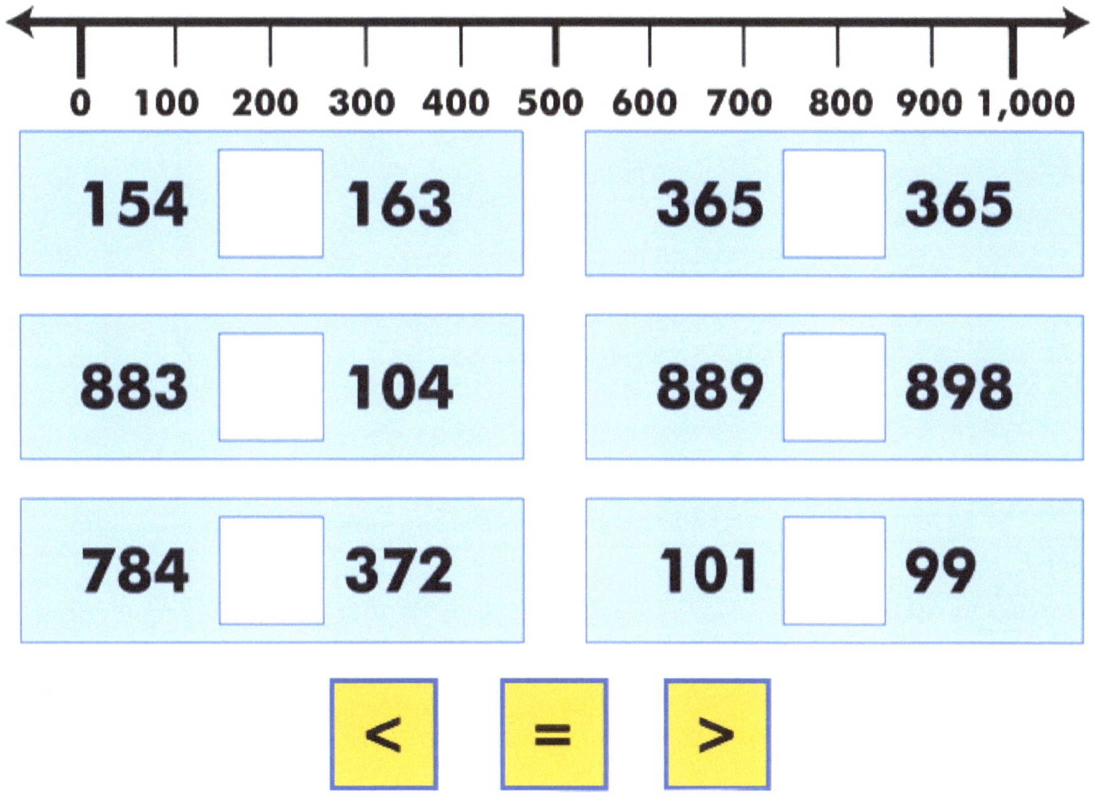

Name____

Compare and Order Numbers Quiz

Circle the correct answer.

(1) **17 < 19   True or false.**

(2) **Which statement is not correct?**
- **(A)** **34 > 21**
- **(B)** **87 < 78**
- **(C)** **123 > 103**
- **(D)** **844 < 944**

(3) **Which number is the smallest? 74   64   91   94**

(4) **Which number is the largest?   145   514   414   235**

Addition & Subtraction

Key Vocabulary

addend

sum

difference

number sentence

Give Matteus the same number of strawberries as Alison.
Hint; first count Alison's strawberries and that will help you to decide how many to add
to Matteus.

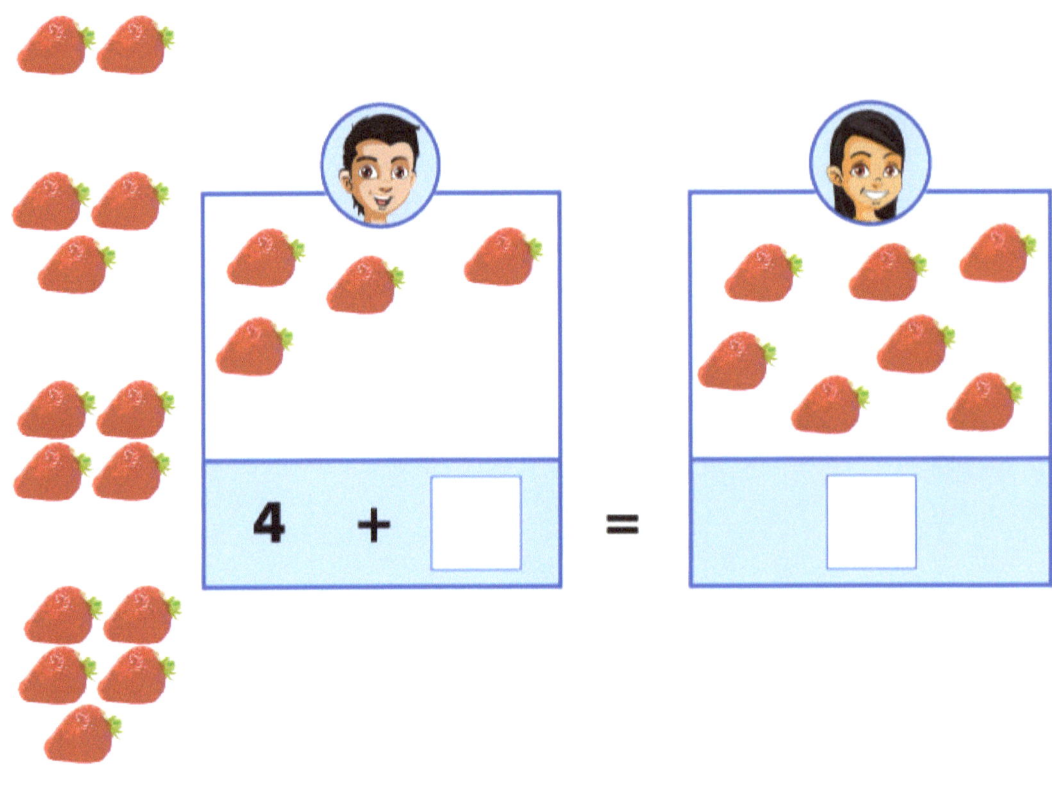

Addition counting on

$$5 + 3 = \boxed{?}$$

Look at the illustrations.

Start with five and
count three more; 6, 7,
8. to get the answer.

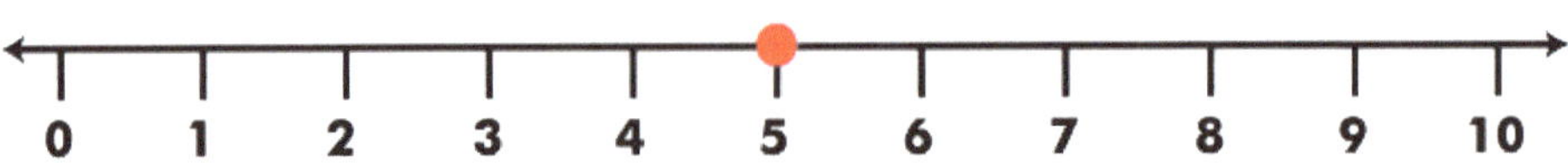

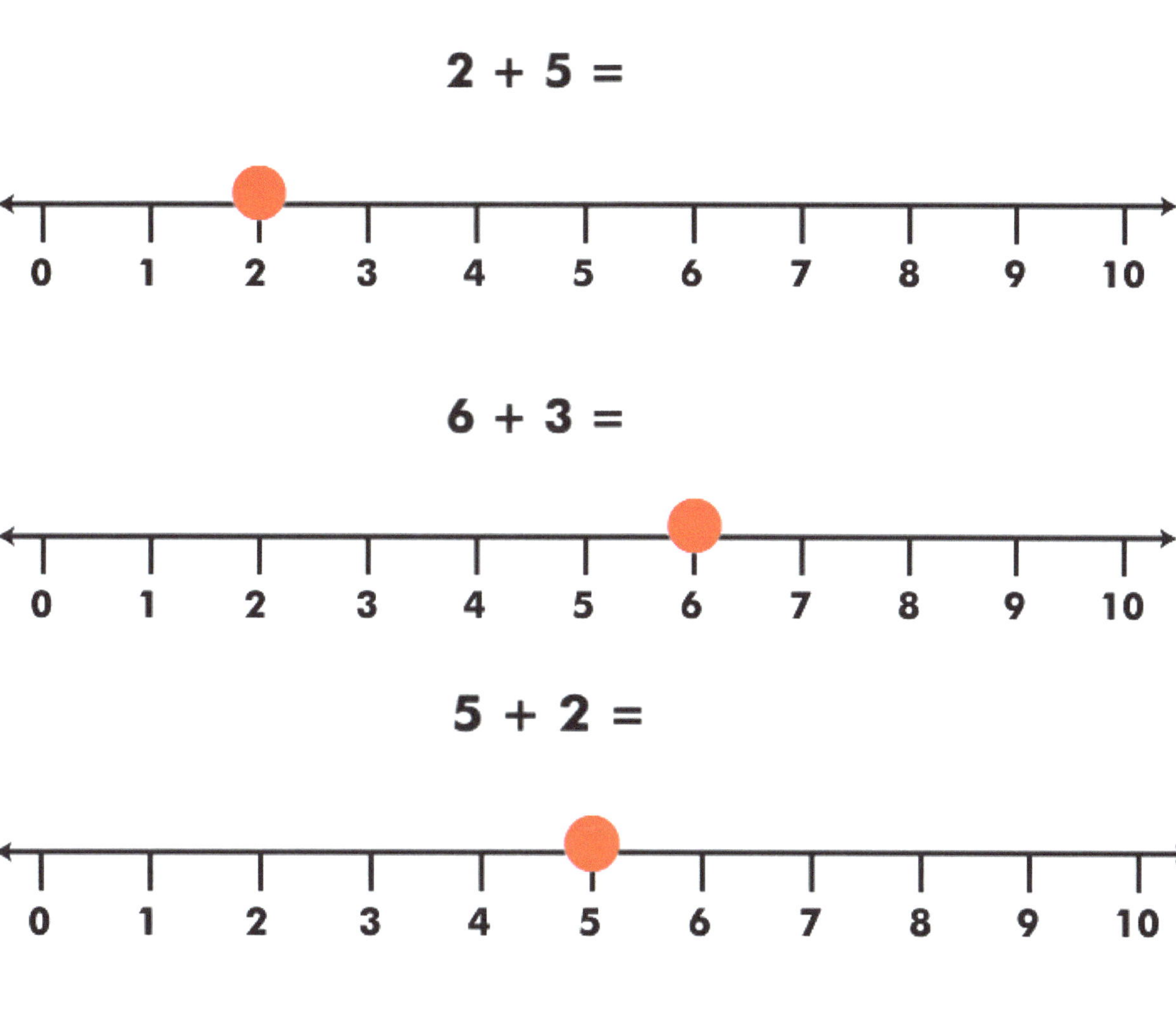

2 + 5 =

6 + 3 =

5 + 2 =

3 + 6 =

Counting back
Look at the illustrations.

Start with eight and count back three; 7, 6, 5, to get the answer.

$$8 - 3 =$$

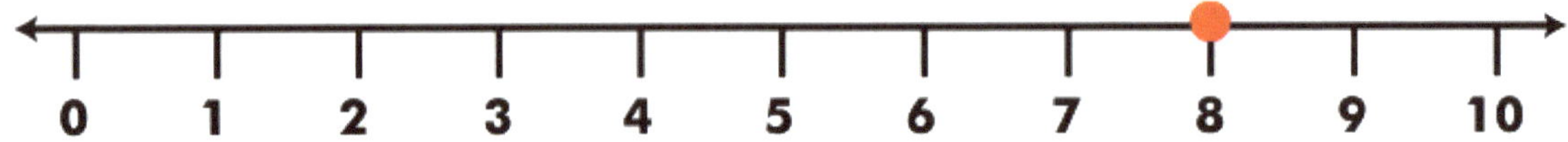

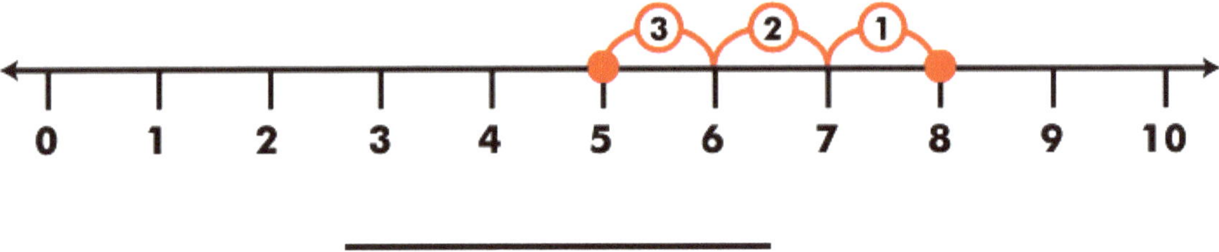

____

Practice counting back

$$10 - 4 =$$

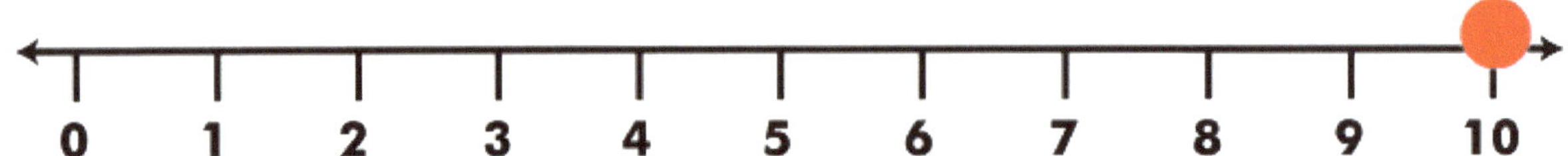

$$5 - 5 =$$

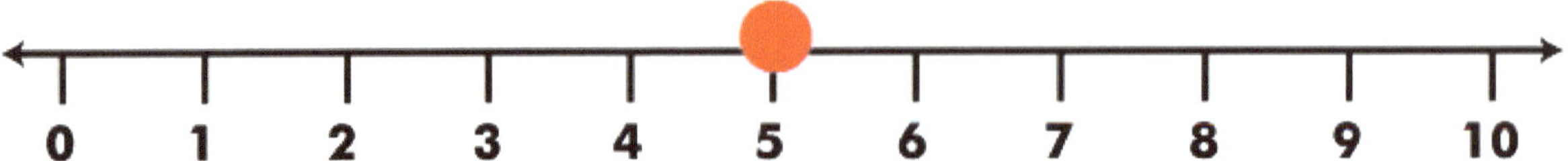

Practice addition and subtraction

1 1 + 4 = ☐   **2** 5 – 1 = ☐

3 7 + 2 = ☐   **4** 9 – 2 = ☐

5 6 + 4 = ☐   **6** 10 – 6 = ☐

───────────

The 4, 5, 9, fact family.

Look at the equations you can make with these three numbers. They are related.

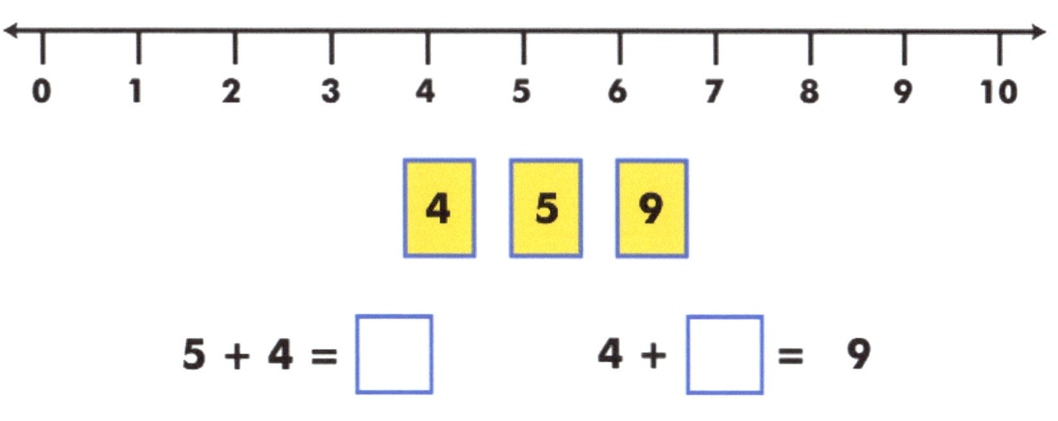

1 Tori has 3 erasers. Owen has 4 erasers.
How many erasers are there altogether?

2 James has 10 cookies. He eats 4 cookies.
How many cookies does he have left?

3 Alicia has $9. She spends $7 on a gift for her sister and
$2 on a soda. How much money does she have left?

Name_________________________________

Addition and Subtraction Quiz

Circle or fill in the correct answer.

 **True or false? 11 − 3 = 8**

 **Which problem equals 12?**

    A   **4 + 5 = ?**

    B   **7 + 5 = ?**

    C   **3 + 8 = ?**

    D   **5 + 6 = ?**

3   **9 − 4 = ?**

4   **What is the missing value? 6 + ? = 9**

Fractions

Key Vocabulary

half

third

fourth

sixth

twelfth

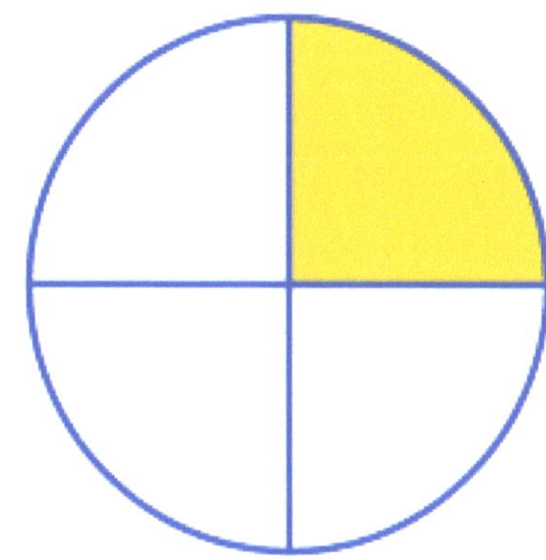

How many parts does this circle have?

How many parts are shaded yellow?

What fraction is shaded yellow? ———— one fourth

How many parts are shaded blue?

What fraction is shaded blue?   two fourths

What fraction is shaded orange?   one fourth

Here you see that the circle has two fourths or
2/4 shaded blue.

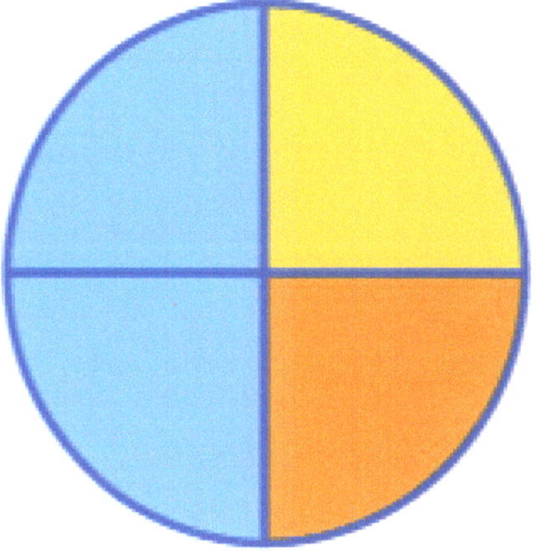

What's another way of writing 2/4?_________

Use the illustration below as a hint.

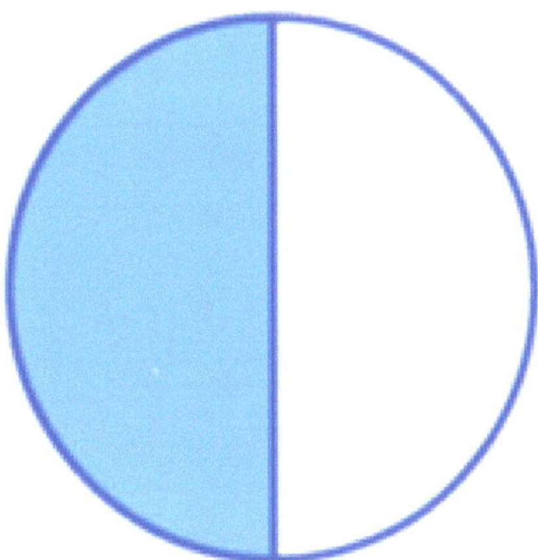

Complete this table.

Is there another way of writing 4/4? __________

Color	# Parts	Fraction
(yellow quarter)	1	$\dfrac{1}{4}$
(two blue quarters)	2	$\dfrac{2}{4}$
(orange quarter)	1	$\dfrac{1}{4}$
Total (circle divided into quarters)		$\dfrac{}{4}$

____

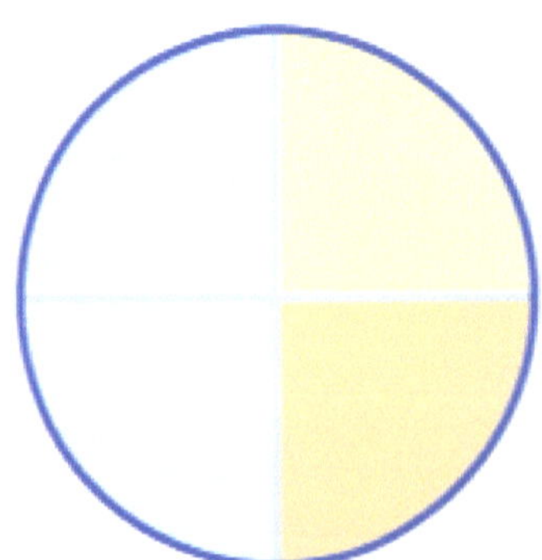

What fraction of each shape is shaded?
Fill in your answers.

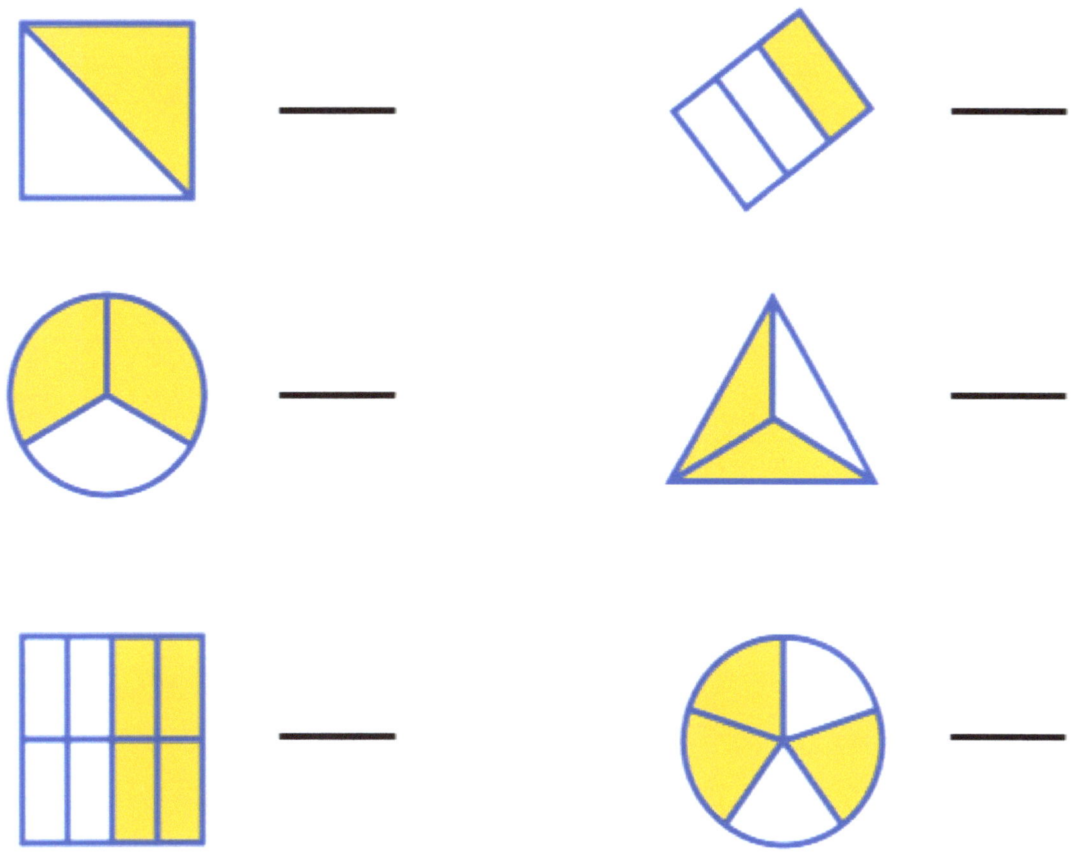

What fraction of each shape is NOT shaded?

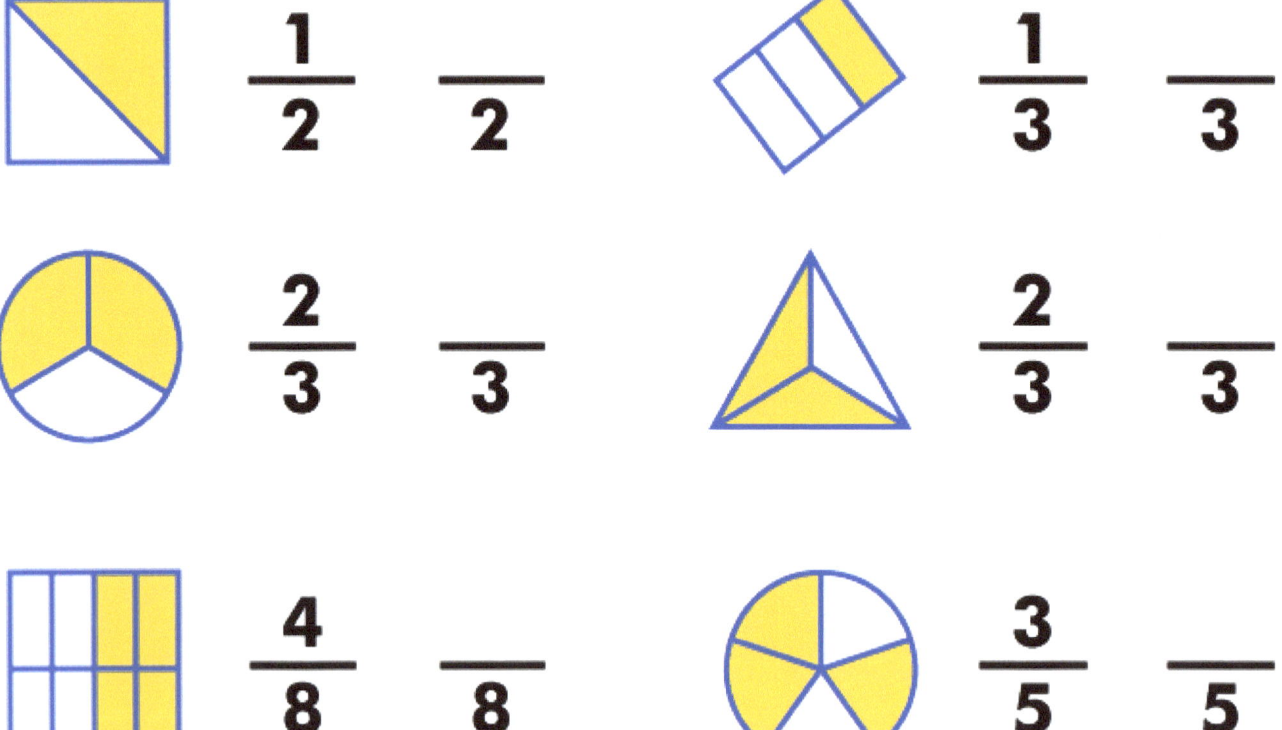

$$\frac{1}{2} \qquad \frac{\ }{2} \qquad\qquad \frac{1}{3} \qquad \frac{\ }{3}$$

$$\frac{2}{3} \qquad \frac{\ }{3} \qquad\qquad \frac{2}{3} \qquad \frac{\ }{3}$$

$$\frac{4}{8} \qquad \frac{\ }{8} \qquad\qquad \frac{3}{5} \qquad \frac{\ }{5}$$

Shade the figure to represent the fraction.

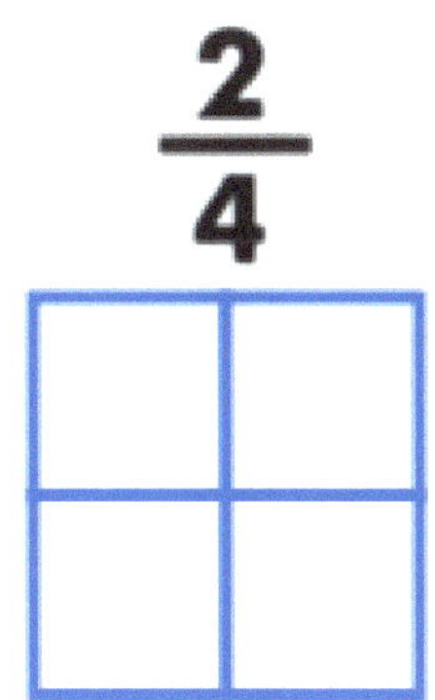

$$\frac{2}{4}$$

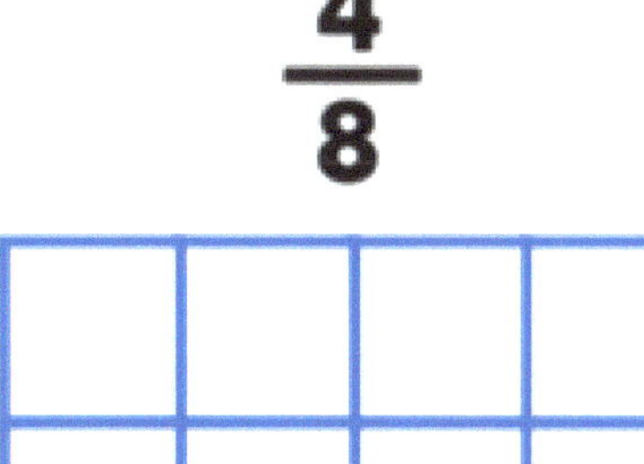

$$\frac{4}{8}$$

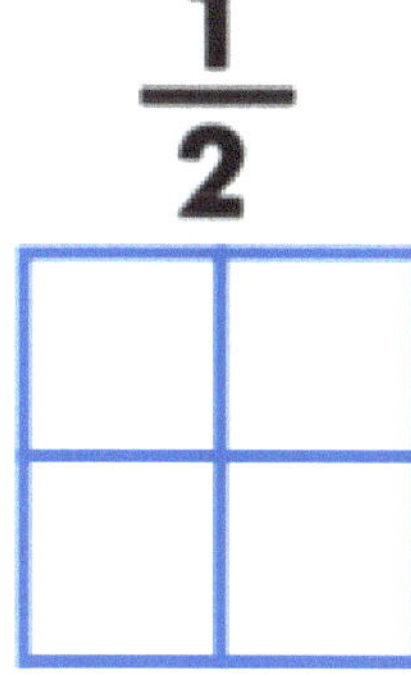

$$\frac{1}{2}$$

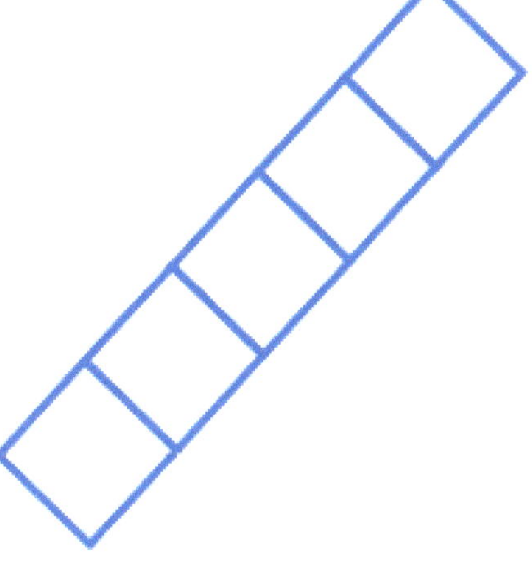

$$\frac{3}{5}$$

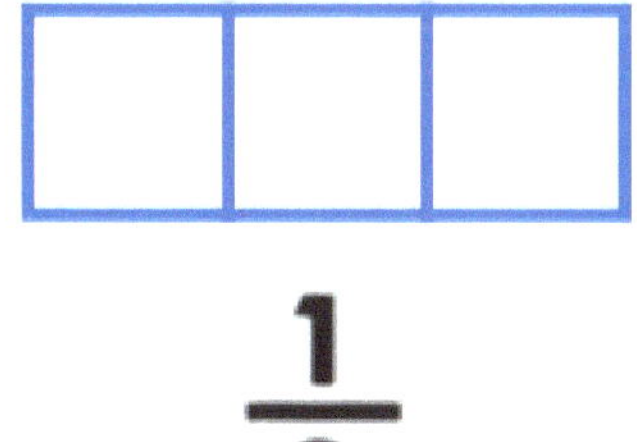

$$\frac{1}{3}$$

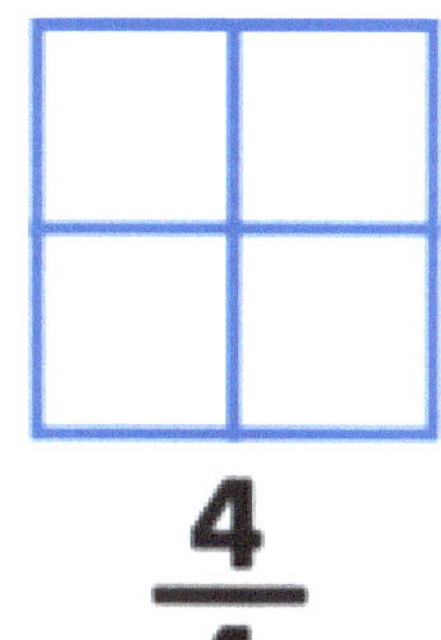

$$\frac{4}{4}$$

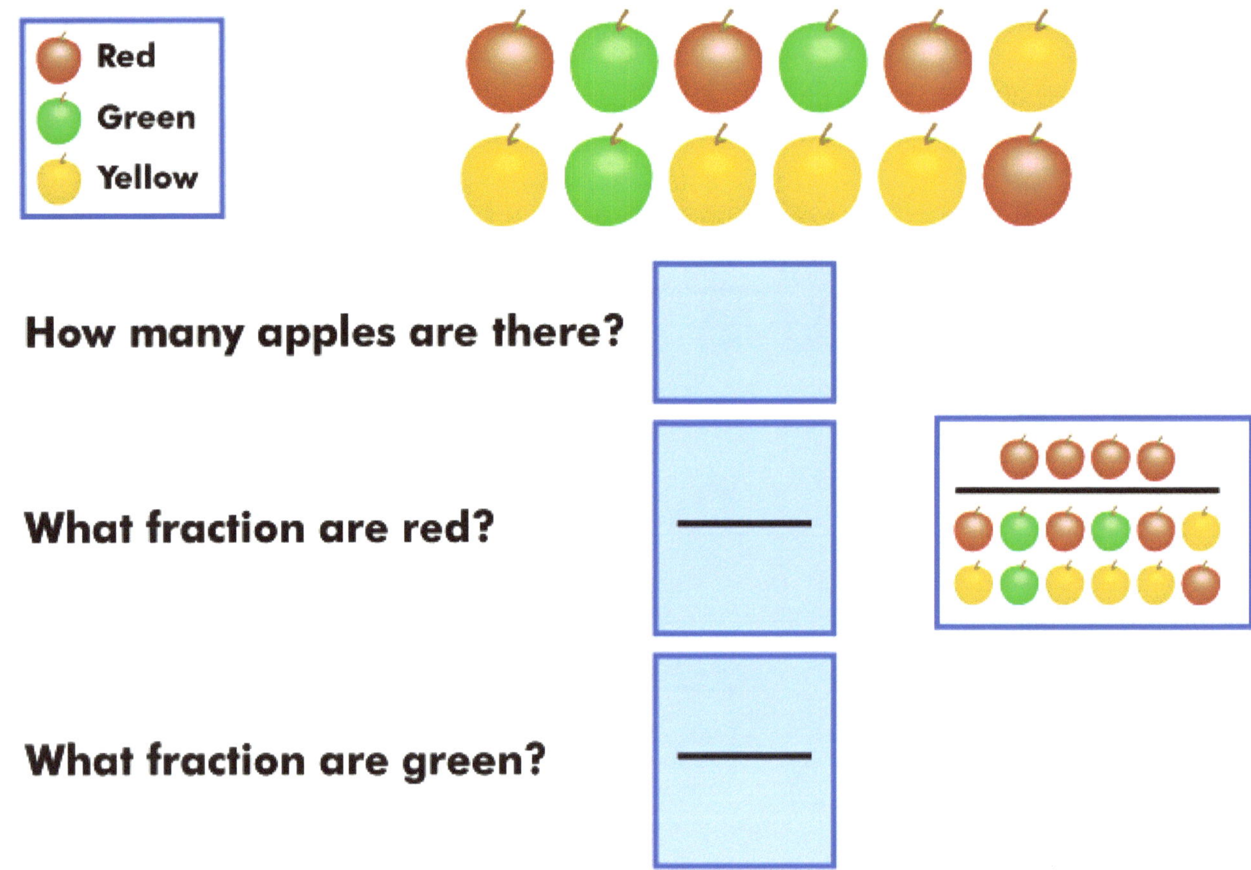

How many apples are there?

What fraction are red?

What fraction are green?

Fractions Quiz

Circle or fill in the correct answer.

 **What fraction of this triangle is shaded?**

 $\frac{1}{2}$  $\frac{1}{3}$  $\frac{1}{4}$  $\frac{1}{6}$ 

 **What fraction of this rectangle is shaded?**

Ⓐ $\frac{3}{4}$ Ⓑ $\frac{1}{3}$ Ⓒ $\frac{2}{3}$ Ⓓ $\frac{3}{3}$ 

 **What fraction of these apples have been bitten?**

 **What fraction of this figure has *not* been shaded?**

Newburyport, MA 01950

1-800-596-3175

OnBoard Academics employs teachers to make lessons for teachers! We create and publish a wide range of aligned lessons in math, science and ELA for use on most EdTech devices including whiteboard, tablets, computers and pdfs for printing.

All of our lessons are aligned to the common core, the Next Generation Science Standards and all state standards.

If you like our products please visit our website for information on individual lessons, teachers licenses, building licenses, district licenses and subscriptions.

Thank you for using OnBoard Academic products.